Welcome to Lapland

Discover the magical Arctic region that stretches across Finland, Sweden, Norway, and even Russia! Here you'll find reindeer, snowy forests, and the dancing Northern Lights, as well as the unique Sami culture.

Every page brings new adventures, fun facts, and surprises – so let's bundle up and explore Lapland together!

Where is Lapland on the map?

Up, up, up in the north!

If you look at a map of Europe and go all the way to the top, you'll find Lapland. But here's a surprise: Lapland is not just in one country—it spreads across four countries: Norway, Sweden, Finland, and Russia.

There's also a very important line running right through Lapland: the Arctic Circle. This is an invisible line around the Earth. When you cross it, something amazing happens:

In some parts of Lapland, at the north of the Arctic Circle, you can play outside at midnight without a flashlight in summer!

🌞 In summer, the sun never sets for weeks (it's called the Midnight Sun).

🌃 In winter, the sun barely rises, and you get long, dark days (that's the Polar Night).

GAME

Lapland is part of 4 countries. Can you circle them on this map? (only part of Russia, as it it huge)

What is Lapland exactly?

A land with no borders—but full of wonders!

Lapland is not a country, it's a region – a big natural and cultural area that spreads across the far north.

How do we know where Lapland starts and ends?

It depends on who you ask:

For geographers, Lapland is the northernmost part of Norway, Sweden, Finland, and Russia's Kola Peninsula.

For the Sámi people (the Indigenous people of the region), Lapland is their homeland, called Sápmi.

For travelers, Lapland often means the land of reindeer, Sámi culture, snowy forests, and northern lights.

So, Lapland doesn't have a strict border line. Instead, it's defined by its nature, people, and traditions. That's what makes it so special!

GAME

What belongs to Lapland?

Lapland is full of unique things. But not everything below belongs there! Circle the features that really describe Lapland.

Sandy deserts with camels

Vast forests and tundra

Ancient castles

Penguins living in icy colonies

Reindeer herding and Sámi traditions

Tropical rainforests and monkeys

Hot beaches with palm trees

Tall pyramids and mummies

The feeling of being in the Far North, where nature rules

Northern Lights (Aurora Borealis)

The Midnight Sun & The Polar Night

The Earth is like a giant spinning top that's leaning to one side. This lean is called the tilt of the Earth's axis. Because of this tilt, different parts of the world get sunlight at different times of the year. In summer, the North Pole tilts toward the sun. This means places north of the Arctic Circle (66°33′ N latitude) can get sunlight all day and all night. That's the Midnight Sun. In winter, the North Pole tilts away from the sun. Then, in the same areas, the sun can't climb above the horizon. That's the Polar Night. So, it's not magic—it's the way our planet is tilted as it travels around the sun!

How long does it last?

Close to the Arctic Circle (like in Rovaniemi, Finland), the Midnight Sun lasts about 1 day in June. The farther north you go, the longer it lasts. At the very top of Lapland, it can last over 70 days without a single sunset! The same is true in winter: the Polar Night lasts just a few days near the Arctic Circle, but up to 2 whole months in the far north.

What does it feel like?

During the Midnight Sun, people sometimes get confused because their body wants to sleep, but the sky is still bright. Locals use thick curtains called blackout curtains to sleep. During the Polar Night, it's not pitch black all the time—there's a magical blue twilight around noon called "the blue moment".

GAME

True or False?

Can you tell what's true about the Midnight Sun and the Polar Night, and what's just a trick?

1. In Lapland's summer, plants grow super fast because they get sunlight day and night.
2. At the South Pole, the seasons are the reverse of Lapland: when Lapland has the Midnight Sun, the South Pole has the Polar Night.
3. The Arctic Circle line always stays in the exact same place and never moves.
4. During Polar Night, Lapland is completely black 24 hours a day, without even a glow in the sky.
5. Some people in Lapland wear sunglasses at midnight in summer because the sun is still shining.
6. The Arctic Circle isn't fixed. It shifts slightly because the Earth wobbles as it spins.

Aurora Borealis (Northern Lights)

The Aurora Borealis is one of Lapland's most magical sights. It happens when particles from the sun crash into the Earth's atmosphere near the North Pole. The sky lights up with glowing ribbons of green, purple, pink, and blue. In Lapland, you can see them almost every clear night from September to March.

Scientists say the best time to see the Northern Lights is between 9 PM and 2 AM on clear, dark nights. You can even use special apps to track the lights and go on your own Northern Lights hunt!

In Lapland, some hotels have cozy glass igloos where you can watch the Northern Lights from your bed–and some even offer a special wake-up call if the lights appear during the night!

GAME

Draw Your Own Northern Lights!

Who are the Sámi?

The Sámi are the Indigenous people of the Far North. They have lived in Sápmi (Lapland) for thousands of years – long before modern borders divided the land into Norway, Sweden, Finland, and Russia.

Today, there are about 80,000–100,000 Sámi people. Most live in Norway (around 40,000), followed by Sweden (20,000), Finland (10,000), and Russia's Kola Peninsula (about 2,000).

GAME Find the missing word!

❄️ = A
⛰️ = N
🦌 = L
🔥 = D
🌲 = P

What Defines Them?

Languages: The Sámi have 9–10 Sámi languages still spoken today.

Reindeer Herding: Not all Sámi herd reindeer, but it's a famous tradition.

Handicrafts (Duodji): Colorful clothes, knives, and jewelry made from natural materials.

Joik: A unique style of singing – like telling a story through sound.

A Journey Through the History of the Lapland

First Settlers (10,000 BC - 3,000 BC)

After the Ice Age, hunter-gatherers settled in Lapland. The ancestors of the Sámi arrived later, at least 5,000 years ago.

Sámi Culture Emerges (3,000 BC - 1,000 AD)

Sámi communities developed reindeer herding, fishing, hunting, animist beliefs, and beautiful handicraft traditions.

Scandinavian Influence Begins (12th - 16th century)

Norwegians and later Swedes began settling near fjords and extending influence over Sámi lands. The first church was built in 1603.

Colonial Pressure & Sami Persecution (1603 - 1700s)

Sámi lands were taxed, missionaries destroyed idols. A Sami man, Lars Nilsson, was executed in 1693 for witchcraft.

Scheffer's Lapponia (1673)

Johannes Scheffer published Lapponia (1673), the first European study of Sámi culture and life.

Border Agreements & Sami Rights (1700s - 1826)

The 1751 Lapp Codicil guaranteed Sámi herding across Sweden-Norway borders. Later treaties (1826) shaped Finland-Norway-Russia borders, disrupting traditional routes.

Industrialization & Conflicts (1800s – mid 1900s)

The expansion of industries such as forestry and mining in the Nordic countries led to increased pressure on traditional Sámi lands and practices.

Lapland Gold Rush (1868 – 1880s)

Gold discovered on the Ivalo River led to a local gold rush and settlements like Kultala in the 1870s.

WWII & Lapland War (1944 – 1945)

WWII's Lapland War devastated towns across Sápmi. Borders shifted again.

Industrial Growth & Decline (1945 – 1990s)

Post-war industrialization brought forestry, mining, hydroelectric plants–but later, economic decline as industries became unprofitable.

Sámi Cultural Revival (1970s – 1996)

Sámi communities across national borders campaigned for rights, leading to grazing laws and, in Finland, the Sámi Parliament in 1996.

Modern Administration (1996 – today)

Sámi representation continues; Finland restructured regions in 2010. Lapland remains a cross-border cultural region–not a unified country–known as Sápmi.

Vikings Meet the Sámi

The Sámi were amazing hunters, reindeer herders, and traders. They spoke their own languages and lived very differently from their Norse (Viking) neighbors. Long before the Viking Age, Roman and medieval writers already described mysterious nomadic peoples in the Far North–these were likely the ancestors of today's Sámi.

The Vikings were famous for sailing ships, trading, and sometimes raiding. Most of the time, they stayed near the coasts. But sometimes, their journeys took them inland–north and east–towards Sápmi (Lapland). They never lived deep inside Lapland, but they did meet and trade with the Sámi.

Signs of Contact

Rock carvings of boats in Northern Sweden, made between 800–1300 suggest Norse traders reached far into Sámi lands.

Sámi people brought reindeer skins, furs, and meat to trade with coastal Norwegians. These trading traditions lasted for hundreds of years!

Some Sámi believed in a thunder god who sounds a lot like Thor, the Viking thunder god with a hammer. In Norse sagas, Vikings often described the Sámi as wise and magical, living close to the spirits of nature.

GAME What's Viking, What's Sámi?

The Vikings traded with the Sámi. Can you spot which things are truly Viking?

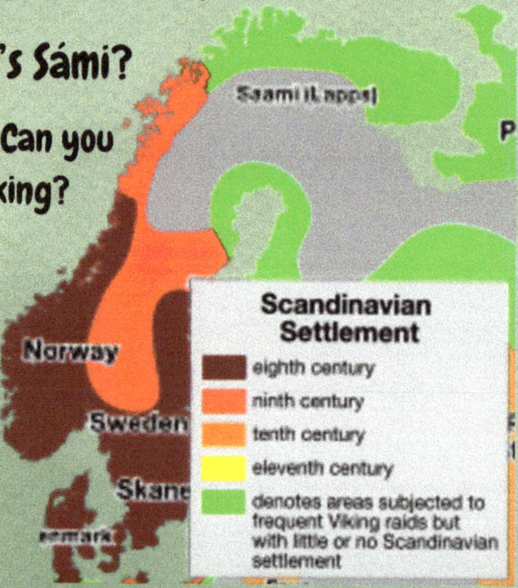

A Tough Time for Sami Children

From the mid-1800s well into the 1900s, governments in Norway (and in somewhat similar ways in Sweden, Finland, and Russia) tried to make the Sámi "more like everyone else." This was called the assimilation policy, or "Norwegianization" in Norway. It meant Sámi children were taken or sent to boarding schools where they were punished if they spoke Sámi. Some even had their hair cut or were teased for how they spoke or dressed. This forced separation lasted for months at a time. Many families stopped teaching Sámi at home, fearing it would hurt their children. By the late 20th century, many Sámi had no language skills in their mother tongue and sometimes didn't even know they were Sámi or felt ashamed of it.

Moving Toward Healing

From the 1960s onward, things began to change for the better:

Norway ended its assimilation laws around the 1960s.

In 1969, Sámi students were legally allowed to be educated in their own language.

Schools, universities, and a Sámi Parliament (established in 1989) helped bring back Sámi culture and language.

In 2024, Norway officially apologized to the Sámi for these policies and began efforts for reconciliation.

The Sámi Flag

A New Flag for an Ancient People

The Sámi flag was officially adopted in 1986. It represents all Sámi people across Norway, Sweden, Finland, and Russia.

Red = fire, warmth, and the sun

Yellow = the sun's light and energy

Green = nature and the forests

Blue = water and the sky

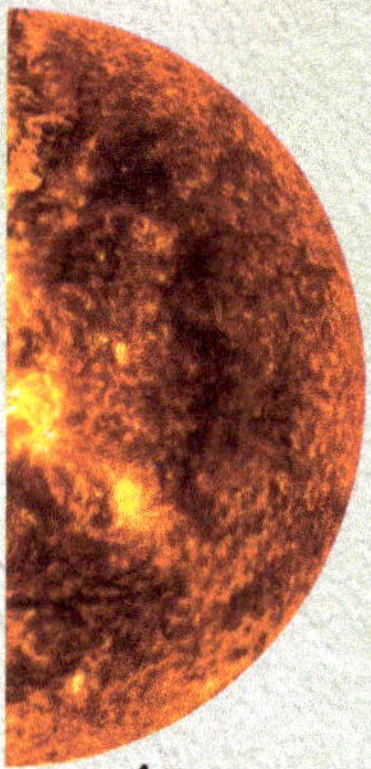

The circle on the flag has two halves:
Red half = the sun, giver of life and energy
Blue half = the moon, important for guiding people at night
This circle was inspired by an old Sámi shaman drum and by the poem "Päiven Parne" ("Sons of the Sun"), which says that the Sámi are children of the sun.

GAME

Color the Sámi Flag!

It's the 6th of February, Sámi National Day! Little Áilu is proudly raising his flag. Can you help him celebrate? Color Áilu and his flag, then make the colors shine bright! Red, blue, green, and yellow—just like the Sámi colors. Don't forget the circle in the middle!

Let's Speak Sámi!

Sámi people don't all speak the same language! There are about 10 different Sámi languages across Norway, Sweden, Finland, and Russia, spoken for thousands of years. Some Sámi languages are so different that speakers might not understand each other. The most common is Northern Sámi, which is used in schools, books, and TV. That's the language we'll learn a few words from today!

GAME

Can You Guess the Sámi Word?

Look at the pictures on the left–what do you think each one is called in Northern Sámi? Draw a line to match it with the words on the right. Take a guess, and learn something new while playing like a Sámi explorer!

Golbma

Beaivi

Ii leat

Máttaráhkká

Nisson

Báze dearvan

Nissonjá

Buerie båeteme

Guokte

Okta

Bassi

Boazu

Giitu

Reindeer – Kings of the North

Reindeer are one of the most famous animals of Lapland! They are not just important for the Sámi people–they are also amazing creatures perfectly built for life in the Arctic. Let's discover their secrets!

Reindeer are great long-distance runners and can travel up to 80 km in a single day.

Nose

Reindeer noses warm icy air before it goes to their lungs and keep heat and moisture in when they breathe out. Like a natural heater!

Antlers

Both males and females grow them. It's the only deer species in the world that does!

Fur

Two layers: soft wool + hollow hairs. Keeps them warm at -40°C and even helps them float in water, just like a life jacket.

Tail

It is short and stubby, which helps prevent frostbite and keeps them warm.

Eyes That Change Color

Golden in summer, blue in winter–helps them see in endless light or darkness.

Hooves

Soft in summer for wet ground, hard and sharp in winter to cut through ice and snow. They can dig through thick snow to reach their favorite food: reindeer lichen.

Ankles

They make a clicking sound when reindeer walk, useful for herders that can hear them in snowstorms!

Reindeer Herding in Lapland

Reindeer herding has been part of Sámi life for hundreds of years. The Sámi people didn't always keep reindeer in large herds. Long ago, they mainly hunted wild reindeer. Over time, they began to domesticate and herd them, turning reindeer into both helpers and companions. Today, reindeer herding is not just about survival—it's also a living tradition, passed down in Sámi families from generation to generation.

Why Reindeer Are So Important

Food: Meat and milk.

Clothing: Warm hides and furs.

Transport: Pulling sleds across snow.

Tools: Antlers and bones were used to make everyday objects.

Herding Today

In Lapland, reindeer roam freely but are marked with special ear-cuts to show which family owns them. Herders use snowmobiles, motorbikes, and drones alongside traditional skis and sleds. Reindeer are gathered in round-ups several times a year.

GAME

Count the Reindeer!

How many reindeer are there in the image below?

Life as a Reindeer Herder

Hi! My name is Niilas. My family are reindeer herders in Lapland, where winters are long and snowy. Reindeer are part of my life every day. Our reindeer roam free most of the year. Each family has its own ear-mark to know which ones are ours. In winter and spring, we gather them in round-ups. Long ago, herders used skis and sleds–my grandfather still remembers! Today we also use snowmobiles or drones, but we keep old traditions too. Reindeer give us food, warm hides, and antlers for tools. They are not just animals–they are part of our family and culture. Life as a herder means being outdoors in every kind of weather. It can be hard, but I love it. The reindeer, the forests, and the sky all feel connected. One day, I want to keep herding, just like my parents and grandparents.

Sámi Reindeer Races

Imagine skiing across snow as fast as the wind... pulled by a reindeer! That's what happens in reindeer races, one of Lapland's most thrilling traditions.

What Is a Reindeer Race?

Reindeer races usually take place in spring, when the snow is still thick but the days are lighter. Racers wear skis and hold onto reins while their reindeer dash across a snowy track–or even on frozen rivers! Sometimes the races happen in cities like Tromsø (Norway) or Rovaniemi (Finland), where whole streets are turned into snowy racecourses.

Why Reindeer Racing?

Reindeer racing is not only fun –it's also a way to celebrate Sámi culture and bring people together. It combines old traditions with excitement and sport. Families, herders, and visitors all gather to cheer for the reindeer. The fastest reindeer can sprint at up to 60 km/h–that's almost as fast as a car in the city!

GAME

Reindeer race names

EQNRS GNNE

Hint: replace each letter with the one that comes after it in the alphabet. (example: A=B)

Juhani takes part in a reindeer race. Like all the participating reindeer, his also got a funny race name. Can you guess what it is? If you were to take part in a reindeer race, what would your reindeer's race name be?

Elks – The Giants of the Forest

Moose (called elks in Europe) are the biggest members of the deer family–and you can meet them in Lapland's forests!

How Big Are They?

Moose are huge–up to 2 meters tall and heavier than a small car! Their antlers can be as wide as a door.

What Do They Eat?

Leaves, twigs, bark, and even water plants. Moose are great swimmers and can munch underwater!

Where Do They Live?

They live in forests, swamps, and by lakes. Moose are shy, but sometimes cross Lapland's roads.

Moose may look calm, but they are strong and fast. In the wild, it's always best to admire them from far away.

Fun Moose Facts!

Their long legs help in deep snow. They have great smell and hearing, but poor eyesight. Males grow and shed antlers each year. Moose calves (babies) can run just a few days after birth.

GAME True or False – Elks

1. Can swim across lakes.
2. Eat fish in the rivers
3. Males grow new antlers every year.
4. Live in the desert.

Brown Bears

Lapland's forests are home to brown bears, one of Europe's largest animals. Even if you rarely see them in the wild, they play an important role in nature.

How Big Are They?

A grown bear can weigh up to 350 kg–as much as five grown-ups together! When standing on two legs, they can be over 2 meters tall.

Winter Sleep

Bears don't really "sleep" all winter. They go into a deep rest called hibernation from October to April, in dens hidden in caves or under tree roots. During this time, they don't eat or drink!

What Do They Eat?

Brown bears are omnivores (eat. everything): berries, nuts, plants, honey, fish, insects and sometimes larger animals, too.

Bear Cubs

Mothers give birth in winter, while hibernating! The tiny cubs (the size of a squirrel at birth) stay warm with their mom until spring.

Fun Facts

Brown bears can run faster than a horse for short distances.
Despite their size, they are great climbers and swimmers.
Bears usually avoid people–they prefer peace and quiet.

Dot to Dot

Connect the dots from 1 to 25!

The Snow Ghosts of Lapland

Arctic foxes are small, clever animals that live in the coldest parts of Lapland. They're sometimes called "snow ghosts" because they are so quiet and hard to spot.

Winter & Summer Coats

In winter, their fur is pure white–perfect camouflage in snow. In summer, their fur turns brown or grey to blend into rocks and tundra.

What Do They Eat?

They munch on lemmings, birds, fish, and berries–whatever they can find. Sometimes they sneak leftovers from other animals.

Fox Families

Foxes live in burrows. In spring, a mother can have up to 14 cubs–a busy den!

A Rare Lapland Sight

Arctic foxes used to be common here. Now they're endangered because winters are warmer and red foxes are moving north.

GAME

Snow Tracks

Look carefully at the snow tracks on the right. Which ones belong to the arctic fox?

Warmest fur of any Arctic animal.
Can survive at -50°C.
Furry paws = snow boots.
Big tails = fluffy blankets.

Musk Ox – The Hairy Giants

Musk oxen look like something from the Ice Age—and that's because they are! These shaggy animals once lived across Lapland during the last Ice Age. Today, a small herd lives in Norway, where they were reintroduced.

Musk oxen have long hair that almost touches the ground. Underneath, they grow qiviut—one of the warmest furs in the world, softer than cashmere!

Height: about 1.5 meters at the shoulder. Weight: up to 400 kg. Both males and females have strong curved horns.

They graze on grasses, mosses, and Arctic plants. Even in winter, they use their hooves to dig through snow for food.

Musk oxen live in groups. When wolves or bears attack, the herd forms a circle with the calves in the middle and the adults' horns facing out. Teamwork keeps them safe!

Musk oxen can survive in -40°C cold. Their name comes from a "musky" smell males make during mating season. They are closer relatives of goats and sheep than of oxen!

GAME

Ice Age Survival

An Ice Age is when much of Earth was covered in ice. The last one ended about 10,000 years ago. Which animals did not survive the last Ice Age?

1.

2.

3.

4.

Lemmings – Tiny but Mighty!

They may be small, but lemmings are Arctic champions! These little rodents live in Lapland's tundra and are super important for nature. Every few years, the number of lemmings explodes—thousands everywhere! Then suddenly, the numbers drop. This cycle keeps nature in balance, because predators depend on them for food. Owls, arctic foxes, and even wolverines love lemmings. Without lemmings, many bigger animals couldn't survive. In winter, lemmings dig tunnels under the snow. There, they stay warm and safe, chewing on grasses, moss, and roots.

ID CARD

- 12–15 cm long
- Never hibernates.
- Makes squeaky sounds to warn each other of danger.

Lemming

❌Myth Buster!❌

You might have heard lemmings throw themselves off cliffs. Not true! That myth came from a movie in the 1950s. Lemmings are smart survivors, not cliff-jumpers.

GAME

Who Eats the Lemming?

Circle the animals that love to eat lemmings!

Kings of the Sky

Even in the far north, the skies are full of life! Some birds stay all year, others visit in summer when the sun never sets.

Snowy Owl – The White Hunter

Big, white, and silent hunters.
Perfectly camouflaged in snow.
Love to eat lemmings–sometimes dozens a day!
Parents guard their chicks fiercely on the tundra.
The snowy owl was made world-famous by Harry Potter (Hedwig!).

Golden Eagle – The Mountain Ruler

Huge wingspan: up to 2.3 meters!
Soars over mountains and forests.
Hunts hares, birds, and even young reindeer.
Seen as a symbol of power and freedom.

Willow Ptarmigan - The Color Changer -

Brown feathers in summer, white in winter.
Masters of camouflage!
Important food for foxes and eagles.

Lapland Longspur & Other Summer Visitors

Many birds migrate thousands of kilometers to nest here in summer. They take advantage of the 24-hour daylight to raise chicks quickly. Lapland becomes a bird paradise when the tundra blooms! Some Arctic terns that nest in Lapland fly all the way to Antarctica each year–the longest journey of any bird.

Arctic Predators

Wolf

Size: Up to 80 kg (as heavy as a grown-up!).
Family life: Wolves live in packs, working together to hunt.
Food: Big animals like reindeer and moose.
Talent: Wolves can run very far and howl to talk to their family.

Wolverine

Size: About 15 kg (like a medium dog).
Lifestyle: Lives alone, roaming huge territories.
Food: Small animals, birds, lemmings, and carrion (dead animals).
Talent: Super strong for its size–can scare animals much bigger than itself!

A wolf is fast: it can run 60 km/h.
A wolverine is nicknamed "the glutton" for its big appetite.
Both are rare and shy–you're lucky if you ever spot one in Lapland!

GAME Who Am I?

Match the clue to the predator:

1. I live in a pack. ➡ ________
2. I sometimes eat leftovers. ➡ ________
3. I howl at night. ➡ ________
4. I'm small but super strong. ➡ ________

Match the clue to the bird:

5. I turn white in winter. ➡ ________
6. I have the biggest wingspan. ➡ ________
7. I eat lemmings all day. ➡ ________

Whales & Seals of Lapland

Lapland isn't only snow, forests, and reindeer–it also has wild coasts on the Arctic Ocean! Here, whales and seals live in icy waters.

Whales

Humpback whales visit northern Norway in winter to feed on herring. Orcas (killer whales) also patrol the fjords. They sing, jump, and splash –giants of the sea!

Seals

Harbor seals and grey seals rest on rocky shores.
In the far north, you might even spot the rare ringed seal, perfectly adapted to ice.
Seals are great swimmers and can hold their breath for over 20 minutes!

GAME

Who Lives Where?

Match the animal to its home:

1.

2.

3.

4.

a.

b.

c.

d.

Huskies – The Snow Dogs

Huskies are not just dogs–they are true Arctic heroes! Strong, smart, and full of energy, they have lived and worked alongside people in the north for thousands of years.

A Bit of History

Huskies were bred by Indigenous peoples of Siberia to pull sleds and help with hunting. They spread to Alaska and Lapland, becoming the ultimate sled dogs. In history, huskies even carried medicine across snowy wilderness to save towns–like during the famous serum run to Nome in 1925!

Husky Facts

Huskies don't get cold easily; their double fur keeps them warm even at –40°C. They can eat huge meals to get energy for running. Their paws act like built-in snow boots. Huskies love to howl together–it sounds like a wolf choir! In summer, huskies often get "vacation time" to rest.

Working Together

Huskies carried explorers across the Arctic.
They helped families travel, hunt, and survive in snowy wilderness.
Today, they still pull sleds–but mostly for sport, tourism, and fun rides!

When you visit Lapland, you can go on a husky safari: ride in a sled, learn dog commands like "Hike!" (go) or "Gee!" (turn right). Some places even let kids help feed or cuddle the dogs.

Types of Huskies

Siberian Husky – fast, with blue or multicolored eyes.

Alaskan Husky – not a breed but a mix, bred to be the best sled dogs.

Malamute – bigger and stronger, perfect for heavy loads.

Life of a Husky

Huskies live in big kennels with other dogs.

A husky team can run for hours in freezing cold.

They love running in teams –working together is their happiest time. Each dog has its own job: leader, swing dog, wheel dog.

GAME

Design Your Husky Sled Team!

Get ready, mushers (sled pulled by huskies driver)! It's your turn to lead a team of huskies through snowy Lapland. Draw your sled and color your huskies. Don't forget to give each husky a fun name–maybe Snowball, Blizzard, or Frosty! Can you create the fastest, fluffiest sled team ever?

Fishing Adventures in Lapland

In Lapland, fishing isn't just a hobby–it's a way of life! With thousands of lakes and rivers, kids and adults can catch fish like pike, perch, salmon, and arctic char. Some lakes are so clear you can see fish swimming deep underwater. Many families have summer cottages by lakes, and fishing is a favorite way to spend time together. Finnish people sometimes cook their catch over an open fire on the shore–fresh and delicious!

In winter, when the lakes freeze over, Finns enjoy ice-hole fishing, cutting a small hole in the ice and patiently waiting for a hungry fish. Ice-hole fishing is sometimes done in complete silence, so the fish aren't scared. Fish like pike and arctic char are very clever–they can spot your bait and even nibble without getting caught!

GAME

Ice-Hole Fishing Maze

Help the little fisherman find his way across the frozen lake to the perfect fishing spot without slipping on the ice.

The 8 Seasons of Lapland

The Sámi follow nature, animals, and the reindeer, not the calendar.
That's why their year has 8 special seasons without fixed dates.
Life, food, and traditions follow the reindeer cycle.

Spring Winter – Gidádálvve

Snow is melting, but nights are still freezing.
Rivers begin to break up with ice.
Families repair tools and prepare for calving.

Spring – Gidá

Reindeer calves are born in May.
Fresh plants start growing, and birds return.
Families move closer to calving grounds.

Early Summer – Gidágiesse

Mosquitoes and insects bother reindeer,
so they move to windy highlands.
People fish in rivers full of salmon.
Berries begin to ripen.

Summer – Giesse

The Midnight Sun never sets—perfect for fishing,
gathering, and visiting.
Reindeer graze in mountains to escape the heat.
Families store fish and berries for the year ahead.

Autumn Summer – Tjakttagiesse

Reindeer are at their strongest.
Sámi families prepare for the autumn roundup.
Mushrooms and berries are still plentiful.

Autumn – Tjaktta

Reindeer are gathered and marked.
Days grow short and cold.
Families store meat, hides, and fat.

Early Winter – Tjakttadálvve

Rivers freeze, the land turns white.
Reindeer are moved to winter pastures.
Families get ready for long travel with sleds.

Winter – Dálvve

The Polar Night brings weeks of darkness.
Reindeer dig through snow for lichen.
Families stay indoors, telling stories, sewing, and carving.

GAME

The Sámi Year Wheel

Fill in the circle calendar with the name of each season, color the slices (for example, blue for winter, green for summer) and draw one small symbol in each slice to show what happens: calf → spring, mosquito → early summer ... and so on!

Beliefs of the Sámi

A Long Time Ago...

The Sámi believed that everything in nature had a spirit–animals, rivers, mountains, even the wind! This is called animism. They honored these spirits to stay safe and healthy and believed in many gods and goddesses.

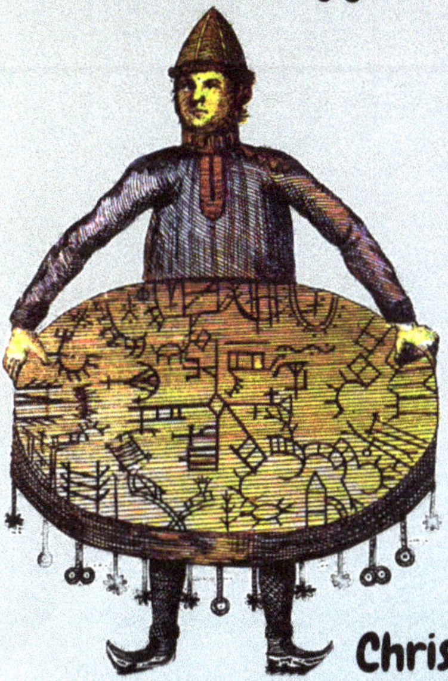

Noaidi – the Shaman

The noaidi, or shaman, had a drum painted with symbols of the sun, moon, animals, and spirits. When he beat the drum, its sound carried him into other worlds. He could speak to spirits, find lost reindeer, or heal the sick. People gathered around as the noaidi's joik rose with the rhythm of the drum. It was both magical and a little scary.

Christian Influence

In the 1600s, missionaries spread Christianity in Lapland. Old traditions were banned, and many sacred drums were destroyed. Today, most Sámi are Christians (Lutheran or Orthodox). But traditional culture lives on: joik singing, respect for nature, and Sámi festivals. Some people still combine old traditions with modern faith. Sámi sacred places (stones, mountains) are still treated with respect.

GAME

Old or New?

Circle which belong to the old religion and which belong to modern times:

1. Noaidi (shaman with a drum)
2. Beaivi, the Sun goddess
3. Joik singing at a festival
4. Going to church on Sunday
5. Horagalles, the thunder god

Sámi Myths & Legends

Imagine it's winter in Lapland. Outside, the snow is deep and the Polar Night makes everything dark. Families gather in the warm glow of a fire. This is when stories come alive! The Sámi told tales about the sun, the moon, animals, and nature spirits. These stories explained the world and taught children how to respect nature.

Beaivi, the Sun Goddess

Beaivi is the goddess of the sun and life. In winter, when the sun vanished for weeks, people worried she was weak and trapped. To help her, they spread butter on reindeer antlers, shining like golden rays. They prayed she would return to warm the Earth. And when she came back, spring began, and joy filled every home.

Máni, the Moon

Children were told: "Don't point at the Moon!" Why? Because Máni, the Moon, might take offense and bring bad luck. Some even said the Moon would come down and steal you away if you teased it too much. (Would you risk pointing?)

Horagalles, the Thunder God

Horagalles was strong and mighty, with a hammer like Thor. He made thunder when he struck the sky, chasing away evil spirits. His rain helped plants grow and filled rivers with fish. But people also feared him –because a careless shout during a storm could make him angry.

The Sacred Bear

The bear was called "forest's old man," and people thought of it as half-human, half-spirit. No one dared say "bear" in the forest–only nicknames like "grandfather." After a hunt, people held a feast to honor the bear's spirit, singing and dancing so it would be reborn. Its bones were carefully buried. This way, the bear would not be offended.

Seita – Sacred Stones

Strange stones and rock formations were thought to hold spirits called seita. Hunters and fishers left offerings there–fish, antlers, or meat–so the spirits would help them. If you forgot to honor a seita, you might return home empty-handed.

Stállo – The Clumsy Giant

Stállo was a huge, foolish giant who loved to bother people. He was greedy and tried to steal food or even children. But he wasn't very clever, and often, Sámi children outsmarted him with tricks. In one tale, a child pretended hot porridge was snow, and Stállo burned his tongue trying to eat it! Brains beat strength.

The Cosmic Reindeer

One story tells of a giant reindeer that carried the sun across the sky on its antlers. Without it, the world would stay dark and frozen. Every morning it ran across the heavens, and every evening it rested. People thanked the reindeer for keeping light alive. It showed how important reindeer were to life itself.

The Invisible People

Legends told of "the hidden people" who lived in the wilderness. They were invisible, but you could sometimes hear them singing or see their footprints in the snow. If treated kindly, they could help you. But if offended, they might steal your belongings or confuse you in the forest. Children were taught never to mock the unseen.

The Path of Souls

The Milky Way was called the "Path of Souls." People believed the dead walked along it to reach the afterlife. On clear nights, families pointed to the bright band of stars and whispered about ancestors traveling there. It was comforting to know their loved ones were never far away.

The Fire Spirit

Fire was alive and sacred. Families never wasted it or spat on it, because that would insult its spirit. Fire gave light in the Polar Night and warmth in the coldest storms. Some believed fire carried prayers upward in its smoke. That's why every flame was treated with care.

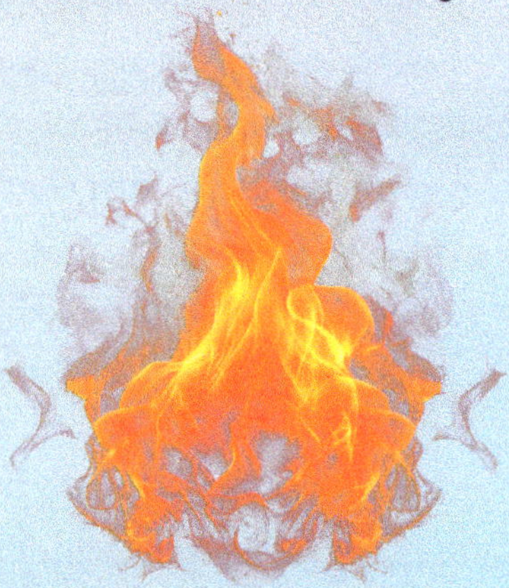

The Spirit of Winter

Winter was not just a season—it was a spirit. People said Winter could freeze your fingers or close your eyes in sleep if you were careless outside. But Winter also gave snow, which protected reindeer pastures and carried sleds across the land. Families both feared and thanked Winter.

The World Tree

The Sámi believed in a great World Tree that connected the heavens, Earth, and underworld. Its branches reached the sky where gods lived. Its roots went deep into hidden realms. Shamans traveled along the tree when they journeyed between worlds. It held all life together, from the smallest lichen to the brightest star.

Wolves and Spirits

Wolves were feared, but respected. Some believed wolves carried messages from the spirit world. A wolf howl was not just noise–it was a sign. If a wolf followed you, it might be a spirit testing your courage. But harming wolves was dangerous, because their spirits would always seek revenge.

The Clever Fox

In Sámi stories, the fox is always tricky and clever. He could fool bears, outwit wolves, and sneak food from people. Children loved fox tales because they were funny and full of surprises. Sometimes the fox's cleverness saved him, but other times it got him into trouble.

GAME True or False – Sámi Myths Edition

1. The bear was called "forest's old man."
2. Sámi shamans used guitars to talk to spirits.
3. Butter was offered to the Sun goddess.
4. Stállo was cleverer than all the humans.
5. Children were told never to point at the Moon.
6. The Sámi believed wolves were just dangerous animals with no spirit power.
7. Seita stones were ordinary rocks used for building houses.
8. The Milky Way was known as the "Path of Souls".
9. Sámi believed fire was sacred and never to be disrespected.
10. Hidden people lived in the wilderness, sometimes helping, sometimes tricking travelers.

Northern Lights Legends

When the green, red, and purple lights dance across Lapland's sky, the Sámi don't just see nature—they see stories. For centuries, the Aurora Borealis was full of mystery, magic, and warnings.

Reindeer Sparks

Some Sámi believed the lights were caused by reindeer running across the sky. Their sharp antlers struck the heavens, sending out showers of sparks that glowed in the night.

The Fox's Tail

Another tale says a magical Arctic fox raced across the snowy mountains. Each time its fluffy tail brushed the snow, it sent sparks flying up into the sky—creating the shimmering lights.

The Souls Above

For many Sámi, the Northern Lights were the souls of the dead, dancing in the heavens. They shone down to remind the living to behave with respect and kindness.

GAME

Circle the animals that appear in Sámi Aurora legends!

Don't Whistle or Point!

Children were warned never to whistle, sing, or point at the Aurora. The lights could swoop down, grab you, or bring bad luck! Instead, people watched quietly, with respect.

Sámi Handicrafts – Duodji

"Duodji" is the Sámi way of making useful things by hand – tools, clothes, jewelry, even sleds! They are made from what nature offers: wood, bone, antler, leather, wool. Every item is both practical and beautiful.

Reindeer Crafts

Antlers become combs, buttons, knife handles. Leather is made into bags, shoes, ropes. Even sinew (tendon) was thread for sewing!

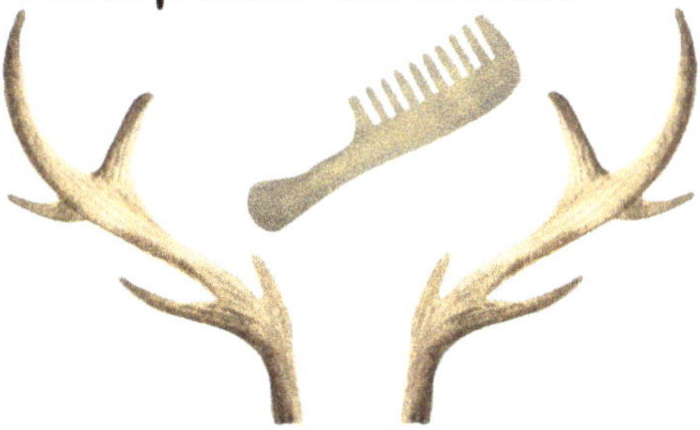

Wooden Mugs (Kuksa or Guksi)

Carved from birch burl (a special round growth on the tree). Always personal–everyone has their own mug! Strong and light, perfect for drinking water from a stream. Often decorated with carvings or reindeer symbols.

Shoes with Grass Socks

Reindeer-skin shoes (nuvttagat) have upturned toes to walk in snow. Filled with dry grass instead of socks– warm and cozy!

Jewelry & Silver Brooches (Risku)

Worn on gákti (traditional clothing). Round shape, often shiny like the sun. Believed to bring protection and good luck.

Knives (Leuku and Puukko)

A Sámi knife isn't just for cutting – it's a tool for daily life: preparing food, fixing sleds, or carving wood. The handle is often made of curly birch, decorated with patterns. The sheath is leather, sometimes with reindeer motifs.

Belts

Colorful wool belts are woven with bright patterns. They're both beautiful and practical: you can attach pouches, knives, or tools to them. The colors and patterns can show where a person comes from.

Beadwork & Embroidery

Sámi clothes (called gákti) often sparkle with beads, tin thread, and embroidery. Patterns aren't random – they tell about family, region, and tradition. Beads were once traded with Norse and Russian merchants, showing how connected Lapland was to the world.

Sámi Sleds (Pulkka)

Small wooden sleds pulled by reindeer. Used for carrying children, food, or tools across snowy land. Still used today in reindeer herding!

A kuksa mug is never supposed to be washed with soap–just rinsed with clean water. Soap would spoil its taste and spirit!

Some duodji patterns are so unique that they work like a "signature" – you can tell who made the object just by looking at the style.

GAME

Design your own Sámi mug or belt!

Traditional Costumes

The Sámi wear a special traditional outfit called gákti. It's colorful, practical, and full of meaning! People wear it during festivals, weddings, reindeer round-ups, and National Sámi Day.

Colors & Patterns

Most gákti use bright colors like: blue, red, yellow, green, and white.

The colors and patterns can show:

- Which region you come from
- Your family or village
- If you're married or not

Accessories

Gákti is often adorned with:

- Belts with tin, silver, or colorful weaving
- Shawls with fringes and embroidery
- Silver brooches and jewelry
- Boots made from reindeer hide

GAME Sámi Hat

The shape of the hat can tell if someone is from Norway, Sweden, Finland, or Russia–and whether they're a man or woman! What would your hat look like?

Made from Nature

Traditional materials include:

- Reindeer leather and fur
- Wool cloth
- Tin thread and beads
- Birch and antler for buttons

Siida Sámi Museum

Siida is a Sámi museum and nature center in Inari, in northern Finnish Lapland, where Arctic nature and Sámi culture come together. The museum includes a large open-air area of about 7 hectares with almost 50 old buildings–such as traditional Sámi homes, barns, and storage huts–that were brought there to protect and preserve them. Siida also includes the Skolt Sámi Heritage House from Sevettijärvi, which represents another Sámi group in Finland. The museum shares the story of Sámi life in the past, present, and future, and shows how closely people and nature are connected. It is also an important place where Sámi people can see their own history and culture presented by their own community. In 2024, Siida was named European Museum of the Year.

If you're heading to Sweden... Visit Ájtte – The Mountain and Sámi Museum in Jokkmokk!

If your journey takes you to Norway... Stop at the Varanger Sámi Museum (Varjjat Sámi Musea) in Varangerbotn!

Sámi Food Traditions

Sámi food comes from nature! For thousands of years, people in Lapland have cooked using what they could gather, fish, hunt or herd. Their meals are simple, warm, full of energy—perfect for Arctic weather!

Reindeer meat is the most traditional Sámi food. It can be: smoked, dried, roasted or cooked in stews.

Arctic waters are full of fish like: salmon, trout, whitefish. Fish can be eaten fresh, smoked, or dried for winter.

In summer, the Sámi pick wild berries such as cloudberries, blueberries, lingonberries. They're eaten fresh or made into jams and sauces.

Even in the Arctic, people found edible plants! They used roots, sorrel, angelica, birch leaves, and mushrooms for soups, teas, and medicine.

Sámi food often had no spices or sugar—the taste came from smoking and fresh ingredients.

Color the Fish!

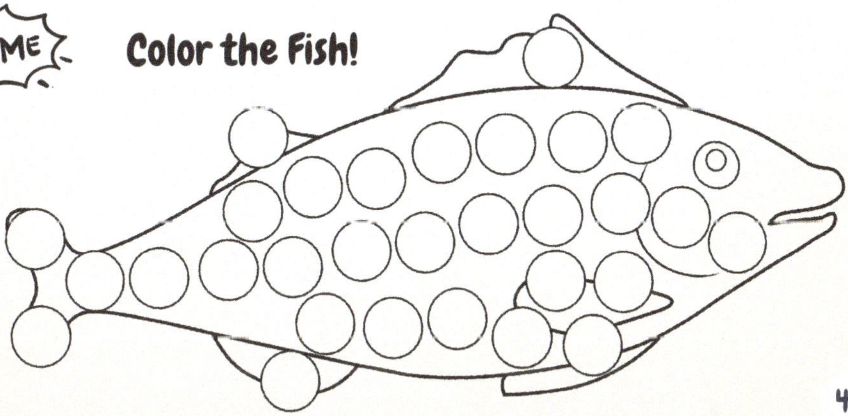

Mushroom Picking

In Lapland, exploring the forest isn't just an adventure – it's a treasure hunt! When autumn arrives, families grab baskets and head into the woods to pick wild mushrooms. The forests around Rovaniemi, Inari, and beyond are full of chanterelles, porcini, and funnel mushrooms. But explorers must be careful–some mushrooms are delicious, and others can make you sick! Thanks to Everyman's Right (Jokamiehenoikeus), anyone can walk in nature and pick mushrooms, even if the land isn't theirs. People in Lapland eat mushrooms in soups, sauces, or dried for winter.

GAME

Name That Mushroom!

1. Bright yellow and shaped like a tiny trumpet. Grows in mossy forests and smells a bit fruity.

2. Has a big brown cap and a thick white stem. Soft like a sponge underneath the cap.

3. Smooth round cap that "leaks" white liquid when cut or broken.

4. Looks wrinkly and brain-shaped. A tricky one–can be poisonous unless prepared properly.

5. The famous red mushroom with white spots. Looks magical–but is poisonous!

6. Poisonous mushroom with a white cap, white gills, and a white stem.

a. Destroying angel

b. Fly agaric

c. Chanterelles

d. Porcini

e. False morels

f. Milk caps

Berry Picking

Lapland's forests are bursting with berries in summer and early autumn! Kids and adults fill buckets with juicy blueberries, bright red lingonberries, and golden cloudberries (also called "Lapland's gold"). These berries are eaten fresh, frozen, turned into jam, or baked in pies. Because of Everyman's Right, anyone in Lapland can go berry picking almost anywhere–on fells, in forests, and even near lakes. It's a tradition, a family activity, and sometimes even a job!

1. Golden-orange berry that grows in bogs. Also called "Lapland's gold".

2. Dark blue berry that stains your fingers and is used in pies and pancakes.

3. Small, bright red berry made into sauces and jams.

4. Tiny black berry that grows close to the ground.

5. Bright red berries on a twiggy plant – birds can eat them, humans cannot!

6. Shiny red berry from a pretty white flower – looks nice but is toxic!

7. Purple-black berry that is dangerous to eat, despite looking juicy.

Name That Berry!

a. Deadly Nightshade

b. Lingonberry

c. Blueberry

d. Crowberry

e. Cloudberry

f. Lily of the Valley

g. Daphne mezereum

Traditional Sámi Dishes

Sámi meals come from nature and are often cooked in simple but clever ways. Here are some real dishes people enjoy in Sápmi:

Suovas = Smoked reindeer meat, often sliced and fried in a pan. It can be eaten in flatbread or with mashed potatoes.

Bidos = A warm reindeer stew made with meat, potatoes, carrots, and water. It's often served at festivals like Sámi National Day.

Kuivaliha = Dried meat, usually reindeer or elk, hung outside in cold, windy weather. It becomes chewy like jerky!

Gáhkku = A soft Sámi flatbread made on a griddle. It can be eaten with butter, meat, or jam.

Smoked Fish = Salmon, trout, or whitefish are smoked over a fire and eaten hot or cold.

Fish Soup = Made with potatoes, onions, and chunks of fresh fish cooked in broth.

Blood Pancakes / Blood Sausages = Made from reindeer blood mixed with flour or grains, then fried or cooked into sausages.

Berries with Cream = Cloudberries, lingonberries, or blueberries served with fresh cream or milk.

Fat Broth (Riebani Baktu) = A hearty soup from boiled reindeer bones and fat–full of energy for cold days.

Food & Ingredients

Match the Sámi Dish to What It's Made Of!

A. Bidos	1. Flour, blood, grains
B. Kuivaliha	2. Cloudberries or blueberries, milk or cream
C. Smoked Fish	3. Slices of dried reindeer or elk meat
D. Blood sausages	4. Fresh trout or salmon, fire or smoke
E. Berries with Cream	5. Reindeer meat, potatoes, carrots

Homes of the Past

Long ago, the Sámi people moved across the snowy lands of Lapland, following their reindeer, rivers, and forests. To live comfortably in this cold, magical region, they built different kinds of homes depending on where they were and how long they stayed.

Lavvu: Cone-shaped tent, similar to a tipi. Made with wooden poles and reindeer hides or woolen fabric. Easy to take down and move.

Turf Huts: Built partly into the ground using stones, wood, and turf. Thick walls kept people warm in freezing winters. Often had just one room with a central fire.

Wooden huts: common by rivers and fjords for fishing, as well as in forests for hunting and trapping.

Snow shelters existed but were not common homes. Hunters sometimes built "luovvi" or snow caves to rest during winter trips—not permanent houses like Inuit igloos.

Village Life Then & Now

Long ago, Sámi families didn't live in big towns—they moved with their reindeer across the snowy forests, lakes, and fjelds. Homes were spread out, and families stayed in small groups. In summer, they lived in one place; in winter, they moved to another.

If you visited a traditional Sámi settlement, you'd see wooden pens full of reindeer, smoke rising from the holes in lavvu tents, food storages raised high on stilts to keep it safe from animals, and wooden sleds and tools lying outside. Life was simple, but clever!

Today, villages in Lapland look very different. Houses are built from wood, brick, or modern logs. You'll find schools, small shops, reindeer fences, and even snowmobiles instead of sleds. Electricity and internet reach even the tiniest villages, but some families still follow old traditions, moving seasonally between summer and winter homes, now using cabins and modern huts.

Life in the Extreme Cold

Lapland can be freezing, with snow covering the land for many months and temperatures dropping far below zero. But Sámi people have lived here for centuries, and they've learned clever ways to stay warm! One of the secrets is clothing. Sámi wear layers of wool, furs, and reindeer skins to trap heat. Thick mittens, scarves, and boots keep fingers and toes cozy, while colorful traditional gákti outfits add style to survival. Homes are just as smart. Reindeer skins were used as soft mattresses, keeping families warm at night. Smoke from fires didn't just cook food–it also helped keep mosquitoes and other bugs away. Doors were low to trap heat inside, and sometimes families even shared their huts with animals to stay extra toasty during icy nights. Every piece of clothing, every part of the home, and every little trick had a purpose. Living in extreme cold isn't just about surviving–it's about being clever, using what nature gives you, and making life cozy, even when it's freezing outside!

Winter Survival in Lapland!

Circle the smart ways to stay warm in the icy Arctic! Some tips are clever, some are silly–can you find them all?

1. Wear layers of wool and furs.
2. Keep doors low to trap heat.
3. Use a hairdryer to melt snow.
4. Share your hut with animals for warmth.
5. Drink ice-cold water only.
6. Sleep on a reindeer-skin mattress.
7. Eat lots of cold snow to stay hydrated.
8. Wear mittens, scarves, and boots.
9. Only wear cotton clothing.
10. Stand outside all night without a coat.

Long Journeys by Sled

A long time ago, before snowmobiles and cars, Sámi people and northern travelers moved across snowy Lapland using sleds drawn by reindeer (and sometimes husky)! These sleds could glide over deep snow, frozen lakes, and wide open fells, carrying families, goods, or even important messages. People traveled for many reasons. Some were moving between summer and winter pastures, following their reindeer herds. Others went to trade, visit neighbors, or deliver supplies. In the early 20th century, husky sleds played a famous role in history: in Lapland and northern Finland, sled teams carried life-saving vaccines to remote villages during harsh winters when roads were impassable! Long journeys on the sled were challenging. Travelers wore layers of wool and furs and sometimes shared the sled with reindeer or dogs for warmth. They learned to read the snow and the stars, find shelter, and recognize frozen rivers and safe paths. Small fires kept food warm, and stops were planned near familiar huts or open shelters. Even children helped guide the reindeer, learning skills that would keep them safe in Lapland's harsh winter.

Help the sled find its way through the snow!

Trading Trails in the Snowy North

Many Sámi families and northern traders followed snowy paths and frozen rivers that stretched across what is now Finland, Sweden, Norway, and Russia. These weren't marked roads—just frozen lakes, reindeer tracks, and secret shortcuts through the forest! Sleds were packed with things the north was famous for: warm reindeer hides, dried fish and meat, antlers and bones used to make tools, handmade cups, belts, and knives. In return, travelers brought back salt (very valuable!), flour and grains, metal for tools, glass beads and fabrics.

GAME

Cool (and Cold!) Facts

Read the sentences and find out which letters each character stands for. Write them in order in the blank spaces to find out a secret word.

Win✫er was ✫he BES✫ ✫ime ✫o ✫ravel—snow made ✫he land smoo✫her ✫han summer mud and mosqui✫oes!

Fozen ives woked like highways—people could glide fo miles without tuning.

No one used p✷per m✷ps—st✷rs, hills, and snow p✷tterns showed the direction.

Re✾ndeer were l✾ke GPS un✾ts—the✾r ✾nst✾nct and memory let them f✾nd the r✾ght d✾rect✾on even better than people!

Some traders spoke three or four ✦anguages to make dea✦s in different vi✦✦ages.

_ _ _ _ _

✷✾✦

Traveling on Skis and Snowshoes

Long before cars or even sleds reached every corner of Lapland, people slipped across the snow using wooden skis and snowshoes. These weren't for fun at first—they were tools for survival!

Skiing the Old Way

The Sámi are believed to be some of the first skiers in the world—over 5,000 years ago! The word ski comes from Old Norse and means "a split piece of wood"! Skis were made from pine or birch and covered in animal skin so they could glide forward but not slide backward. Hunters and herders used one long ski and one short ski—or even a ski and a pole—to move across hills and forests.

Walking on Snow Without Sinking

Snowshoes worked like giant tennis rackets for your feet. They spread your weight so you didn't disappear into deep snow. People wore them while herding reindeer, hunting, or collecting firewood. They were inspired by animal paws. Wolves, lynxes, and wolverines have wide paws that work a bit like natural snowshoes!

GAME

Who went to the cabin?

From Reindeer to "Vroom!"

As time passed, people in Lapland traded silent skis for something louder and faster: snowmobiles! Snowmobiles began to appear in Lapland in the 1960s. At first, herders didn't trust them... until they realized they could follow big herds much faster. Today, snowmobiles are used for reindeer herding, carrying supplies to cabins, search and rescue, tourism and safaris. Some forest paths have snowmobile traffic signs, like tiny snowy highways. A snowmobile can ride over ice, hills, and frozen swamps where cars can't go. Reindeer still don't love the noise—but many herders use both animals and engines.

GAME

Ols vs. New

You're carrying food, tools, antlers across deep snow, and the wind is howling. You have to travel 40 km across Lapland. Would you pick: skis, snowshoes, a sled with reindeer, or a snowmobile? Write your answer and why below. (Just remember... snowmobiles don't come with fur!)

Sámi Games & Children's Play

Even in the freezing Arctic, Sámi children always found time to play. Their games weren't just for fun—they helped them learn skills they'd need later in life, like balance, throwing, tracking, and working with animals. Sámi toys were often handmade from reindeer bones, wood, antlers, or scraps of fur. Some looked like tiny sleds, animals, or people in gákti clothing!

Reindeer Lasso Practice

Kids learned to throw a suohpan (lasso) by practicing on sticks or horns placed in the snow. They'd compete to see who could catch the "reindeer" from the farthest distance!

Footprints in the Snow

One child made animal tracks using hands or sticks, and the others had to guess the animal—reindeer, fox, or ptarmigan!

Snow Hide-and-Seek

Kids hid behind snowbanks or inside snow tunnels, and the seeker followed footprints to find them.

Summer Play

When the snow melted, children played by rivers and lakes, skipped stones, built tiny boats, and collected colorful stones or feathers.

GAME

Count the Snowballs!

Speaking for the Sámi People

The Sámi live in Finland, Norway, Sweden, and parts of Russia. To protect their culture, languages, and rights, each country created a Sámi Parliament—a special group of representatives elected every 4 years by the Sámi people, who speak for them.

Finland: The Sámi Parliament in Inari helps protect Sámi languages, culture, and rights.

Sweden: The parliament in Kiruna focuses on cultural programs, language preservation, and local development.

Norway: The largest Sámi Parliament, located in Karasjok, deals with culture, education, and land use.

Russia: There is no official parliament, but Sámi councils exist to represent local Sámi communities.

Sámi Parliaments help the Sámi:

Keep languages alive – many children can learn Northern Sámi in school.

Protect the land – decisions about reindeer herding, fishing, and hunting go through them.

Share traditions – music, crafts, and festivals are supported, bringing Sámi from different countries together.

GAME

Imagine This!

If you were a young Sámi representative, what ideas would you bring to your parliament? Protecting reindeer? Saving your language? Or making a giant ice festival?

The Norwegian Sámi Parliament building has a modern design inspired by Sámi tents (lavvu).

Lapland's Taiga – Endless Forests

Imagine walking into a forest so big it feels like it never ends! That's the taiga in Lapland. Here, tall pine, spruce, and birch trees stretch as far as your eyes can see, creating a magical world full of life.

What It Looks Like

Towering trees with green needles or shiny leaves.
Mossy floors, dotted with berries and mushrooms.
Streams, rivers, and small lakes hidden among the trees.

Who Lives Here?

Elk (moose) wandering quietly.
Brown bears napping in the shade or fishing in streams.
Wolves, lynx, and foxes stalking through the undergrowth.
Birds like owls, woodpeckers, and capercaillie singing or pecking around.

The taiga is a berry paradise: blueberries, lingonberries, and cloudberries grow in summer.

Trees in the taiga help the planet by storing huge amounts of carbon in their trunks and soil.

Sámi reindeer herders use taiga forests for winter pastures.

GAME

What Would You See in the Taiga?

Circle all the things you might really see in the taiga!

Tundra – Life on the Bare Hills

High up in northern Lapland, above the tree line, where it's too cold for tall trees to grow, stretches the tundra–a cold, windy landscape where the ground is mostly frozen for much of the year. It may look empty at first, but it's full of tiny plants, amazing animals, and surprising colors during the short summer.

What It Looks Like

Low bushes, mosses, and tiny flowers hugging the ground.
Snow covers most of the land for much of the year.
Rocky hills and wide-open spaces give you a 360° view of the sky.

Who Lives Here?

Reindeer grazing on lichen and moss.
Arctic foxes hunting small rodents.
Birds like ptarmigans and snow buntings nesting in summer.

Lichen is called "reindeer salad" because it's their favorite food!

Plants grow low to the ground to hide from icy winds.

In summer, the tundra blooms with tiny, colorful flowers, like nature's own paintbrush.

GAME

What Would You See in the Tundra?

Circle all the things you might really see in the tundra!

Frozen Ground All Year!

Imagine the ground so cold that it stays frozen all year, even in summer! That's permafrost, a special feature of northern Lapland.

What Is It?

Permafrost is soil, rocks, or sediment that stays below 0°C for at least 2 years. Only the very top layer thaws in summer; deeper layers stay frozen like giant ice cubes under the ground. Permafrost can be hundreds of meters deep.

Life in Permafrost

Arctic plants survive by growing close to the ground and spreading shallow roots. Tiny creatures like worms and microbes hibernate deep underground. Old bones, seeds, and even ancient tools can stay preserved for thousands of years!

Permafrost helps lock carbon in the ground, which affects the Earth's climate.

Building houses on permafrost is tricky: the ground can shift and crack when it melts.

GAME Frozen Treasures in the Permafrost!

Z	F	Y	O	I	A	W	L	T	E	E	T	H	G	F	V	K	E	W	T
V	O	F	E	N	P	A	L	K	O	S	G	B	Y	E	L	R	X	O	W
G	S	U	N	C	Y	O	E	J	Z	Y	Q	O	Q	A	F	K	N	O	L
E	S	R	O	Z	S	O	C	C	Q	S	L	M	T	T	W	A	R	D	Y
Q	I	Q	B	A	N	T	L	E	R	Y	E	V	F	H	P	M	N	Z	H
C	L	O	T	H	I	N	G	H	D	Y	T	E	U	E	P	E	A	B	K
B	U	G	R	U	R	M	U	L	T	P	Q	L	D	R	C	Q	H	M	T
V	R	C	E	R	A	M	I	C	S	S	H	E	L	L	N	P	T	X	O
L	F	N	V	A	C	T	J	D	V	V	T	L	V	C	J	B	U	O	O
Z	F	W	B	K	P	R	W	H	S	V	K	R	X	N	U	Q	S	U	L

- bone
- seed
- fossil
- tool
- clothing
- ceramics
- fur
- teeth
- shell
- wood
- antler
- feather

Glacial Eras – Lapland Under Ice!

A long, long time ago, Lapland looked very different. The region went through several ice ages, with glaciers covering the land, followed by warmer periods called interglacials, when the ice melted and plants and animals could return. Scientists have identified at least four major glacial cycles over the past 2 million years!

During the Glacial Eras

Ice sheets up to 2 kilometers thick covered Lapland. Rivers and lakes were frozen, and most plants could not survive. Only cold-adapted animals like reindeer, arctic foxes and lemmings could live at the edges of the ice. Humans had to move south or live in small ice-free refuges.

During Interglacial Periods

The climate warmed, glaciers retreated, and forests and tundra returned. Plants grew again–shrubs, mosses, and grasses spread across the land. Reindeer herds increased, and early humans returned to hunt and gather.

The Last Ice Age

The last major ice age ended about 11,700 years ago, when Lapland's ice melted and the landscape began to look more like it does today. Meltwater formed lakes, rivers, and fjords, and forests slowly returned to the lower areas. Rocks with scratches (striations) show the path of moving ice.

Ice Age or Warmer Days?

Draw a line from each item to the time it belongs to:

Glacial Era
(very cold, ice everywhere)

Interglacial Era (warmer, forests and animals return)

Mountains & Fells of Lapland

Lapland doesn't have tall, pointy mountains like the Alps or the Rockies. Instead, it's home to fells: wide, treeless hills or mountains with a flat or rounded top. Most of Lapland's fells rise above the tree line, where it's too cold and windy for tall trees to grow. That's why the tops are covered with rocks, moss, and tiny Arctic plants.

What Is Up There?

Above the tree line, the landscape changes into tundra-like terrain:

- ✅ Reindeer moss
- ✅ Dwarf birch
- ✅ Lingonberries & crowberries
- ✅ Lichens and hardy grasses
- ✅ Snow patches that last into summer

What Makes Them Special?

You can see reindeer herds climbing the slopes.

In winter, they turn into ski resorts and sledding areas.

In summer, people hike under the Midnight Sun.

Many fells are important in Sámi stories and traditions.

Long ago, some fells in Lapland were even taller than the Alps! But during the ice ages, giant glaciers covered the land. These massive sheets of ice scraped and crushed the ground, shaving off the tops of the fells and leaving them smooth and round. The glaciers were so heavy that they pushed the land down. When the ice finally melted, the ground began to rise back up—and it's still slowly rising today!

GAME

Lapland Math

A fell used to be 3,800 meters tall. After glaciers scraped it down, it's now 1,500 meters. How many meters were removed? If the land rises about 2 centimeter per year. In how many years will it get to 1,502 meters?

Fjords of Northern Norway

Northern Norway is famous for its fjords—deep, narrow inlets where the sea cuts into the land. Imagine long rivers of salty water snaking between towering cliffs and snowy mountains. These fjords were carved over thousands of years by mighty glaciers during the ice ages. When the ice melted, the sea filled the deep valleys, creating the magical fjords we see today. The fjords can be hundreds of meters deep, deeper than some of the tallest mountains! Some fjords are over 100 km long—long enough to hide entire towns along their shores. Fjords make great hiking adventures: trails often climb to cliffs with breathtaking views of the water below.

Who Lives in the Fjords?

Seals bask on rocky shores.

Sea eagles soar above the cliffs.

Fish like salmon, cod, and herring swim in the waters.

Occasionally, playful porpoises and whales can be spotted!

GAME

Fjord Word Scramble

Can you unscramble the letters to find 10 things related to a fjord?

LAES, LMOSAN, TBOA, CILFF, LRIEGAC, GIEVALL, GLAEE, ROPOPSIE, LAWEH, LEVLAY

North Cape & the Arctic Ocean

Imagine standing at the very top of Europe, where cliffs rise high above the icy Arctic Ocean. This is North Cape, in Norway's Lapland! From here, you can see water stretching farther than your eyes can follow–sometimes all the way to the horizon.

North Cape is often called "the top of the world" because it's the northernmost point of Europe you can reach by car.

There's a visitor center with exhibits about Arctic nature, polar explorers, and Sámi culture.

In summer, the Midnight Sun shines 24 hours a day–you can watch the sun float across the sky without ever setting!

In winter, the Northern Lights dance above the icy ocean–a magical sky show!

The cliffs rise 307 meters (1,007 feet) above sea level, making it a dramatic viewpoint.

Brave adventurers sometimes spot whales, puffins, and seals in the Arctic waters below.

GAME

Word Search: Explore the Arctic Ocean!

W	O	X	D	X	Y	Y	E	H	D	R	O	J	F	D
W	Y	A	Q	S	M	J	P	P	L	Z	B	W	S	N
E	P	U	F	F	I	N	N	N	I	P	H	H	T	U
J	A	X	X	T	H	M	T	O	O	X	D	A	H	S
B	I	N	X	O	S	M	L	R	R	T	W	L	G	T
O	C	U	W	O	N	S	E	T	E	O	H	E	I	H
A	E	I	C	V	O	S	N	H	X	G	C	D	L	G
T	B	C	L	C	G	L	G	C	K	G	L	B	N	I
I	E	D	S	O	I	G	L	A	C	I	E	R	R	N
W	R	U	R	N	Q	T	D	P	Q	X	I	H	E	D
R	G	S	T	W	R	N	C	E	L	C	L	Z	H	I
E	X	P	L	O	R	E	R	R	R	A	U	X	T	M
N	X	P	O	L	A	R	B	E	A	R	E	J	R	E
T	B	M	N	A	E	C	O	H	G	V	V	S	O	J
Y	W	F	F	I	L	C	K	Q	F	T	Z	B	N	G

Fjord
Glacier
Snow
Seal
Whale
Puffin

Boat
Arctic
Ocean
Cliff
Iceberg
Explorer

Polar bear
Northern Lights
Midnight Sun
North Cape

The Amazing Rock Art of Alta

A long, long time ago—6,000 to 7,000 years ago—people lived in what is now northern Norway. They didn't have cameras, phones, or even paper. Instead, they told stories on rocks! These people carved over 6,000 images into the cliffs along the Alta River, and today, these carvings are called the Alta Rock Carvings. They are like a time machine, letting us peek into the lives of people from thousands of years ago. They show how humans lived, hunted, and played in the Arctic long before modern technology.

What You Can See

Reindeer and elk – hunted for food, clothing, and tools.

Boats and fishermen – showing the importance of the sea.

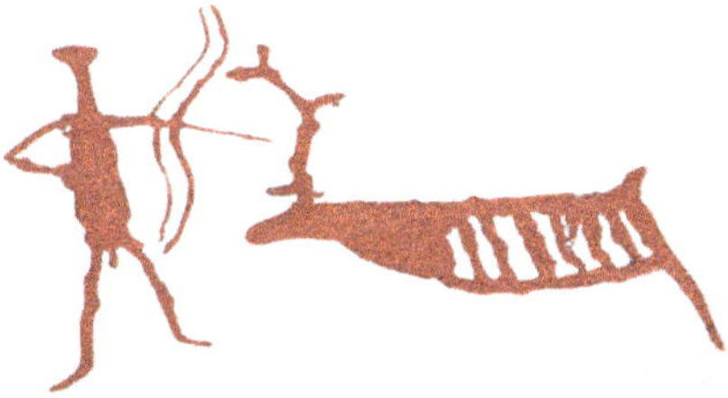

Hunters with bows – keeping the community fed.

Dancing figures – maybe celebrations, rituals, or magical dances!

They show that people back then loved stories and art, just like we do today.

The carvings were made by pecking or scratching the rock with stones.

The carvings are so important that UNESCO declared them a World Heritage Site in 1985.

If you go to Alta, you can walk along a special trail to see the carvings. There's also a museum where you can touch replicas, try carving with safe tools, and learn more about the people who lived here long ago.

Look at pictures of Alta carvings and match each to its meaning

a. reindeer
b. boat
c. hunter
d. fish

Kola Peninsula

Far in the northeast of Europe lies the Kola Peninsula, a special corner of Sámi land and Lapland. While it shares snowy forests, fells, and rivers with Norway, Sweden, and Finland, Kola is truly one-of-a-kind! The Sámi here live in small, remote villages and follow traditions unique to this region. Their clothing, language, and crafts are a little different from Sámi in other countries, making Kola culture very special. The land itself is wild and beautiful, with tundra, deep forests, and rugged mountains. Some areas are almost untouched by humans, giving it a real "wilderness feel". Scientists visit Kola to study the climate, Arctic plants, and wildlife because it is one of the most untouched Arctic regions in Europe. Kola also has some of the oldest rocks in Europe and is rich in minerals–some even glow in the dark! Reindeer roam freely across the land, Arctic foxes sneak through the tundra, and seabirds like puffins nest along the icy coasts. Even rare plants survive here, specially adapted to the harsh Arctic conditions. The peninsula's lakes and fjords are full of fish, making fishing an important way of life for the people who live there, even if some rivers in Kola freeze for over half the year.

GAME

Arctic Story Time

Write a short story about an adventure in Kola, spotting reindeer, birds, or the Northern Lights.

Underground Treasures in Lapland

Even though Lapland looks like endless forests, tundra, and snow, there's a lot hiding underground! The Arctic lands are full of valuable resources that people use in everyday life.

GAME

What Resources Are Found in Lapland?

Draw a line from each resource on the left to what people use it for on the right.

1. Iron ore – especially in Kiruna and Gällivare (Sweden)

2. Copper – found in Sweden, Norway, Finland, and the Kola Peninsula

3. Nickel and cobalt – mainly in Finland and Russia's Kola region

4. Gold and silver – mined in Finland, Norway, and Sweden

5. Platinum – found in northern Norway and Russia

6. Amethyst – found in Luosto, Finland

7. Limestone and granite – quarried in Finnish and Swedish Lapland

A. Jewelry and old coins

B. Roads and buildings

C. Wires and electronics

D. Car and phone batteries

E. Steel for machines and structures

F. Decorations and gemstones

G. Medical equipment and jewelry

Kiruna, in Swedish Lapland, has one of the world's biggest iron mines. It's so huge that the ground beneath the town started to crack and sink. So, the entire town had to move! Houses and shops were taken apart and rebuilt about 3 kilometers away. Some buildings, including a giant wooden church, were lifted and transported in one piece! Imagine your whole neighborhood packing up because of hidden iron –that's Kiruna! One of the biggest town moves in modern history!

Nature, Mining & Debate

Mining is one of the ways people get metals, minerals, and jobs, but it also changes the land and affects those who live there. In Lapland, huge machines dig deep into the earth, sometimes cutting apart mountains or creating long tunnels underground. Rocks are crushed, metals are separated, and the leftover stones are piled nearby. Many of these mines are near Sámi communities and the paths where reindeer travel. When the land is dug up, reindeer struggle to find the grass they need to eat, and the noise from roads and machines can scare them away. Sacred places and old trails may also be disturbed, which is why Sámi people speak up to protect their land and traditions. Nature itself feels the impact. Forests are cut down, rivers and lakes can become polluted, and animals like bears, foxes, wolverines, and birds may have to find new homes. Today, there are rules that make mining safer and cleaner, and people try to replant forests, clean water, recycle metals, and listen to local communities before starting new mines.

GAME

Arctic Future – What Would You Do?

The Arctic is changing! Should new mines open? Who should decide? How can we protect nature and still get what we need?

Mineral Mining for Kids

Lapland is a treasure chest of sparkling gems and shiny metals! Here are some fun, kid-friendly mining activities where you can dig, discover, and learn:

Amethyst Mining at Lampivaara, Luosto

In the heart of Pyhä-Luosto National Park, Lampivaara Amethyst Mine offers a unique experience. Kids can dig for their own amethyst gemstones using traditional tools. The mine operates sustainably, ensuring the environment remains protected. After digging, visitors can keep their "lucky stone" as a souvenir.

Gold Panning at Tankavaara

In Tankavaara, there's the Goldmining Museum where you can learn all about gold, old tools, and famous Finnish gold discoveries. Close to the museum, there are shallow rivers where you can step in, hold a pan, and swirl sand and water to see if you can spot golden flakes–just like a real gold prospector! Adults and guides help you, so it's safe, fun, and exciting. If you don't get to Tankavaara, you can also try your luck at Apukka Resort in Rovaniemi.

Iron Mining at LKAB Visitor Centre, Kiruna

In Kiruna, the LKAB Visitor Centre offers an interactive experience where kids can learn about the region's rich mining history. Guided tours take visitors 540 meters underground to explore the mining tunnels and see the massive machines up close.

Castles of Ice & Snow

Deep in the Arctic, when winter comes and snow blankets the land, magical palaces of ice begin to appear. These are carved from snow and ice, built fresh each year by artists, sculptors, and... dreamers.

Icehotel – Jukkasjärvi, Sweden

The world's first ice hotel! Artists carve bedrooms, ice chandeliers, chairs, beds, and even cups for the ice bar. In spring it melts back into the river.

SnowCastle – Kemi, Finland

A giant snow castle with slides, an ice restaurant, and a hotel. Some parts now stay cold all year so visitors can explore even in summer.

Kirkenes Snowhotel – Norway

Every room is a work of art – carved animals, Sámi stories, and glowing ice sculptures. Guests sleep on reindeer furs in warm sleeping bags! They also have an "365" version so ice rooms are available almost all year!

How They Are Built

First, blocks of ice are cut from rivers (like the Torne River for Jukkasjärvi).

Snow is packed and carved to form walls, floors, ceilings, arches, and sculptures.

Sculptors use chisels and saws to carve details – animals, patterns, towers, even ice chandeliers.

Every year, new artists are invited to propose designs.

When spring melts set in, the ice returns to water – just like it always does.

The rooms stay around -4°C (25°F) all the time.

Ice hotels have ice bars – glasses made of ice!

Artists from around the world design new rooms each year.

GAME

Design Your Own Ice Room

Draw or describe a hotel room made of ice – what would you carve into the walls? A dragon? A dolphin? A secret tunnel?

From Snow to Steam

Imagine stepping from icy snow... into a tiny wooden room as warm as summer! That would be a sauna, one of Lapland's oldest traditions. In Lapland, people have used saunas for over 2,000 years. Back then, they weren't just for relaxing – they were used for: washing (before bathrooms existed!), healing with steam and herbs, giving birth, because it was the cleanest place in the village. Today, saunas are still part of everyday life. Families even have them in their homes and cottages!

Different Kinds of Saunas

Smoke Sauna (Savusauna)

No chimney – smoke fills the room while the fire heats stones. After the fire goes out, people air it out and go in. The walls turn black, but the steam is soft and amazing!

Wood-Burning Sauna

Crackling fire, wooden benches, and that toasty pine smell.

Electric Sauna

Press a button – steam time!

Ice and Snow Saunas

Yep, real! Walls of snow, benches of ice, and heat from a stove inside.

Sauna Rules

The classic routine goes like this:

1. Sit in the hot sauna and toss water on the stones – pssssshh!
2. When you're super warm... RUN OUTSIDE!
3. Roll in the snow or jump in an icy lake
4. Laugh, scream, or shout "AAAH!"
5. Go back in and heat up again!

Kids do it too – with adults nearby of course!

What's with the Tree Whacking?

People sometimes use a "vihta" or "vasta" – a bundle of birch twigs – to gently tap their skin. It sounds silly, but it improves blood flow, smells like a forest and keeps mosquitoes away in summer! Think of it as the world's weirdest massage.

Sit on the wooden benches
Wear swimsuits or towels in public saunas
Stay quiet or tell stories
Drink water or berry juice – never soda!

Color the sauna scene, then test your imagination!

Warm up your creativity by coloring the sauna scene below! When you're done, try the bonus challenge: list 5 things that would MELT in a sauna.

Celebrating in Lapland

In Lapland, life is shaped by light and darkness, snow and sun. That's why people celebrate with festivals that are as unique as the land!

Midnight Sun Festivals

Imagine playing soccer, eating ice cream, or even running a race... at midnight, while the sun is still shining! In Lapland's summer, the sun never sets for weeks. People celebrate with bonfires, music, night markets, and games under the glowing sky.

Polar Night Celebrations

In winter, the opposite happens: the sun disappears for weeks. But Lapland turns this into magic! Villages "encourage" the return of the sun with lantern parades, glowing ice sculptures, fire shows, and warm cocoa. And while waiting, the Northern Lights often dance across the dark skies!

Sámi Easter Festival (Kautokeino, Norway)

Easter isn't just about chocolate eggs here–it's reindeer races, lasso-throwing contests, and colorful Sámi clothing. Families gather to enjoy joik (traditional singing), feasts, and even friendly competitions. It's one of the most exciting Sámi gatherings of the year!

Riddu Riddu Festival (Norway's Arctic Coast)

Every summer, a small village by the sea fills with drums, singing, and color. Sámi people get to meet friends from other Arctic and Indigenous cultures–like Inuit, Native American, and Siberian groups. Children join workshops, paint, sing, and listen to magical stories from around the world.

Sámi National Day (February 6th)

On this day, the Sámi flag is raised across Lapland, and people wear gákti (traditional Sámi clothes) with pride. There are parades, joik performances, reindeer rides, and craft markets. Families cook Sámi food like reindeer stew and flatbread, sharing traditions that go back hundreds of years.

The Sámi National Day marks the first Sámi congress in 1917, when Sámi from Norway and Sweden met to defend their rights. Today, children celebrate too–reminding everyone that Sámi culture is alive and strong!

Lapland Festival Crossword Adventure!

Grab your pencil and see if you can match each clue to the correct word. From reindeer races to glowing Northern Lights, every answer brings Lapland's magic to life!

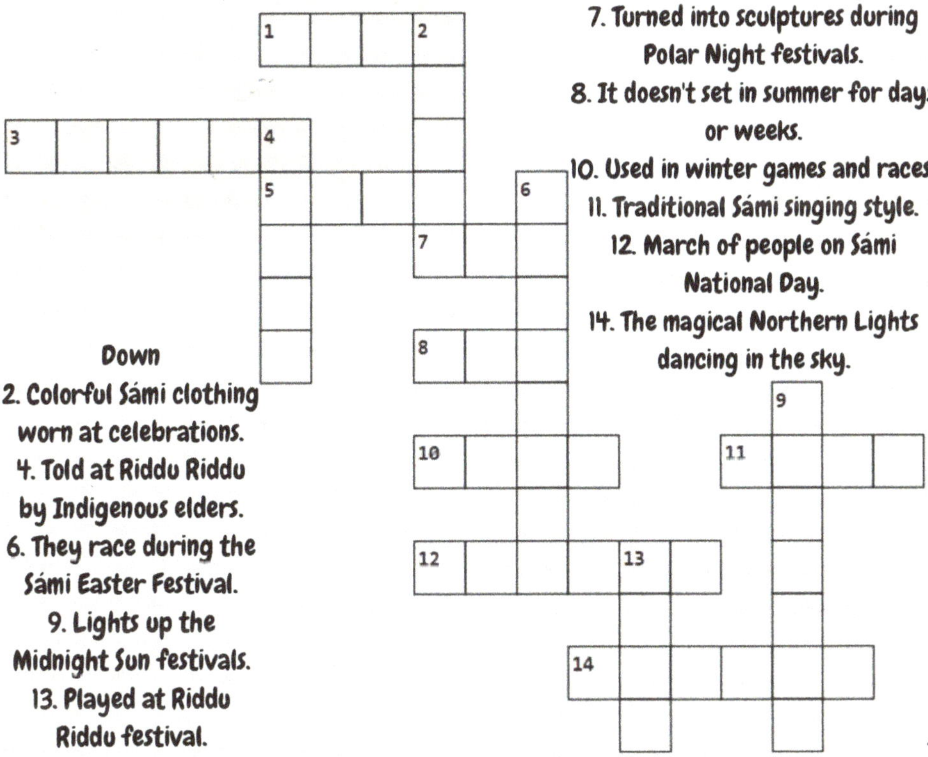

Across

1. Raised on Sámi National Day.
3. Handmade items sold at festival markets.
5. Children sleep in these during summer festivals.
7. Turned into sculptures during Polar Night festivals.
8. It doesn't set in summer for days or weeks.
10. Used in winter games and races.
11. Traditional Sámi singing style.
12. March of people on Sámi National Day.
14. The magical Northern Lights dancing in the sky.

Down

2. Colorful Sámi clothing worn at celebrations.
4. Told at Riddu Riddu by Indigenous elders.
6. They race during the Sámi Easter Festival.
9. Lights up the Midnight Sun festivals.
13. Played at Riddu Riddu festival.

Christmas Traditions in Lapland

Christmas in Lapland is a magical mix of snow, lights, and tasty treats! Families gather to celebrate with joulupöytä, a table full of favorites like baked ham, rice porridge (with a hidden almond for luck!), smoked salmon, and berry jams. In Sámi homes, bidos, a hearty reindeer stew, is a favorite, often eaten with flatbread and sweet cloudberries. Homes sparkle with candles, straw ornaments, and pine decorations. Children build snowmen, make snow angels, and sometimes even take part in small sled races outside. In villages, fires burn to keep people warm, and the scent of gingerbread and cinnamon fills the air.

During the long, dark winters, Christmas lights brighten the snowy forests and make Lapland feel like a fairy tale.

GAME

Santa Claus Flags Match

Santa isn't called the same everywhere in Lapland!

Joulupukki – Finland

Julenissen – Norway

Jultomten – Sweden

Ded Moroz ("Grandfather Frost") –Russia

Write the right Santa name above each country flag!

He also may wear different costumes. In some places, Santa's coat has fur trim, in others it has bright colors or traditional patterns. Even the elves' clothes change depending on which Santa they work for!

Santa's Village in Rovaniemi

Santa's Village is right on the Arctic Circle, and it's magical all year round! Here, children from all over the world come to meet Santa, peek inside his office, bake cookies with Mrs. Santa, and see the elves busily preparing gifts. Reindeer roam nearby, and you can cross the Arctic Circle line–an important spot marked by signs and decorations. The village has shops, cozy cafés, and workshops where children can try small crafts. In winter, ice sculptures and slides add a sparkling, icy wonder. In summer, the sun never sets, making the village feel like a bright, endless playground.

GAME

Spot the Characters!

Santa Claus
Edgar the Polar Bear
Bailey the Husky
Frosty the Snowman
Rudolph the Reindeer
Christmas Carolers
Ailwy the Elf
Skipper the Squirrel
Woody the Woodpecker

Santa's Mailroom

Millions of letters arrive at Santa's Mailroom every year! Santa receives letters in languages from every continent–including some from kids as far away as Australia and Japan! The elves sort them by country, language, and type of wish. Some letters have drawings, others have secret codes. Special letters are delivered directly to Santa, who reads them carefully to plan Christmas Eve. Elves use magical stamps, ribbons, and little carts to move letters around. Children often wonder how Santa remembers everyone–but the elves say it's teamwork and a little Christmas magic!

GAME

Help Julian write his letter to Santa!

Dear Santa,

Please, very, very much, talk to my ___  ***___ and ___***  ***___ to convince yourself that I am a good ___***  ***___.***

If you want to know how I study and what grades I have at ___  ***___, then talk to the ___***  ***___.***

If you want to know how I behave with the ___  ***___ on my street, then talk to all the ___***  ***___.***

But first and foremost you have to talk to ___  ***___. Well, it would be best to talk only with ___***  ***___, because she stays at home all day. You don't have to walk too far to find her. And you, dear Santa Claus, are an old man and can't walk around too much asking about me.***

I am waiting for you with ___  ***___ and impatience!***
Julian

Elf Workshops

Deep in snowy Lapland, elves are busy all year long–not just at Christmas! Their workshops are cozy, colorful places filled with jingling bells, warm lights, and the smell of gingerbread.

Toy-Making Elves build wooden trains, tiny animals, and colorful puzzles.

Letter Reader Elves sort mail sent from children all around the world!

Reindeer Elves polish sleigh bells and pack snacks made of lichen and berries.

Gift-Wrapping Elves twist ribbons, tie bows, and wrap surprises at lightning speed.

Cookie Elves bake gingerbread, berry buns, and cinnamon treats that fill the air with sweet smells.

Ice Artist Elves carve snow dragons, castles, and glowing sculptures out of frozen blocks.

Psst... Some elves have whispering pockets where they keep whispers from the wind, messages from reindeer, and wishes from children. If you could slip one wish into an elf's pocket, what would it be? Write or draw it here!

GAME

Arctic Adventures & Santa Parks

Lapland is also a playground of Arctic adventures! Children can ride husky sleds across snowy forests, explore giant ice sculptures, or zoom down snowy slopes on tubes. Reindeer shows, snowball games, and winter playgrounds make every visit exciting. During the polar night, when the sky is dark, the Northern Lights dance overhead in colorful ribbons. In the summer, the Midnight Sun means you can play and explore without ever seeing darkness. Arctic parks even offer ice fishing, snowmobile rides, and small winter obstacle courses made just for kids. And Santa magic isn't only in Santa's Village—there are other amazing Christmas parks across Lapland:

SantaPark Underground in Rovaniemi
Santa's Cabin in Levi
Elves' Hideaway in Kittilä
Santa's Resort Kakslauttanen in Saariselkä

GAME

What station did the train arrive at?

Answers

Page 2:

Page 3: Vast forests and tundra, Reindeer herding and Sámi traditions, The feeling of being in the Far North, where nature rules, Northern Lights (Aurora Borealis)

Page 5: ✅✅❌❌✅✅

Page 7: Lapland

Page 10:

Vikings 

Sami 

Page 13: Boy = Nisson, Girl = Nissonjá, Sun = Beaivi, Moon = Máttaráhkká, Tree = Bassi, Reindeer = Boazu, Hello = Buerie båeteme, Please = Giitu, Goodbye = Báze dearvan, One = Okta, Two = Guokte, Three = Golbma, Sorry = Ii leat

Page 15: 7, Page 17: FROST HOOF, Page 18: 1.✅, 2.❌, 3.✅, 4.❌

Page 19:

Page 20:

Page 21: 2 – Woolly mammoth, 4 – Sabre-toothed cats

Page 22: Snowy owl, Arctic fox, Wolverine, Ermine

Page 24: 1. Wolf, 2. Wolverine, 3. Wolf, 4. Wolverine, 5. Willow ptarmigan, 6. Golden eagle, 7. Snowy owl

Page 25: 1-d, 2-b, 3-a, 4-c, Page 28:

Page 31: 1, 2, 5 = old / 3, 4 = modern,

Page 35: 1.✅, 2.❌, 3.✅, 4.❌, 5. ✅, 6.❌, 7.❌, 8.✅, 9.✅, 10.✅

Page 36: 1 (reindeer), 3 (fox)

Page 42: 1-c, 2-d. 3-f, 4-e, 5-b, 6-a

Page 43: 1-e, 2-c, 3-b, 4-d, 5-g, 6-f, 7-a, Page 44: A-5, B-3, C-4, D-1, E-2

Page 47: 1.✅, 2.✅, 3.❌, 4.✅, 5.❌, 6.✅, 7.❌, 8.✅, 9.❌, 10.❌

Page 48:

Page 49: TRAIL, Page 50: Anna

Page 52: 12

Page 54:

Page 55:

Page 56:

Z	F	Y	O	I	A	W	L	T	E	E	T	H	G	F	V	K	E	W	T
V	O	F	E	N	P	A	L	K	O	S	G	B	Y	E	L	R	X	O	W
G	S	U	N	C	Y	O	E	J	Z	Y	Q	O	Q	A	F	K	N	O	L
E	S	R	O	Z	S	O	C	C	Q	S	L	M	T	T	W	A	R	D	Y
Q	I	Q	B	A	N	T	L	E	R	Y	E	V	F	H	P	M	N	Z	H
C	L	O	T	H	I	N	G	H	D	Y	T	E	U	E	P	E	A	B	K
B	U	G	R	U	R	M	U	L	T	P	Q	L	D	R	C	Q	H	M	T
V	R	C	E	R	A	M	I	C	S	S	H	E	L	L	N	P	T	X	O
L	F	N	V	A	C	T	J	D	V	V	T	L	V	C	J	B	U	O	O
Z	F	W	B	K	P	R	W	H	S	V	K	R	X	N	U	Q	S	U	L

Page 57:
Glacial era: mammoth, reindeer, arctic fox
Interglacial era: pine trees, birch tree, lake, frog. reindeer, arctic fox

Page 58: 3.800-1,500=2,300
1.502-1,500=2m=200cm; 200:2=100

Page 59: SEAL, SALMON, BOAT, CLIFF, GLACIER, VILLAGE, EAGLE, PORPOISE, WHALE, VALLEY

Page 60:

W	O	X	D	X	Y	Y	E	H	D	R	O	J	F	D
W	Y	A	Q	S	M	J	P	P	L	Z	B	W	S	N
E	P	U	F	F	I	N	N	N	I	P	H	H	T	U
J	A	X	X	T	H	M	T	O	O	X	D	A	H	S
B	I	N	X	O	S	M	L	R	R	T	W	L	G	T
O	C	U	W	O	N	S	E	T	E	O	H	E	I	H
A	E	I	C	V	O	S	N	H	X	G	C	D	L	G
T	B	C	L	C	G	L	G	C	K	G	L	B	N	I
I	E	D	S	O	I	G	L	A	C	I	E	R	R	N
W	R	U	R	N	Q	T	D	P	Q	X	I	H	E	D
R	G	S	T	W	R	N	C	E	L	C	L	Z	H	I
E	X	P	L	O	R	E	R	R	R	A	U	X	T	M
N	X	P	O	L	A	R	B	E	A	R	E	J	R	E
T	B	M	N	A	E	C	O	H	G	V	V	S	O	J
Y	W	F	F	I	L	C	K	Q	F	T	Z	B	N	G

Page 61: 1-c, 2-d. 3-a, 4-b and c (the hunter is on a boat, probably seal hunting)

Page 63: 1-E, 2-C, 3-D, 4-A, 5-G, 6-F, 7-B

Page 69: Chocolate bar, Ice cream, Snowballs. Crayons, Candles, Butter, Wax figures, Cheese, Lip balm Marshmallows

Page71:

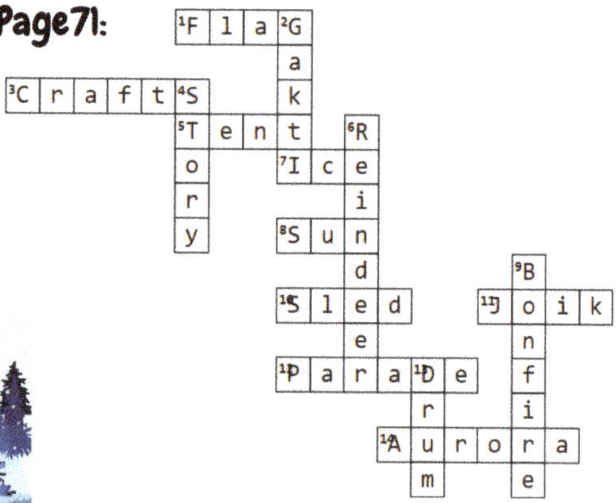

Page 72: Jultomten, Ded Moroz, Julenissen, Joulupukki

Page 73:

Page 74: mother, father, boy, school, teacher, children, neighbors, grandma, grandma, love

Page 76: NORTH POLE

Lapland Travel Map

Our Lapland Journey Comes to an End... for Now!

You've travelled across snowy fells, chased the Northern Lights, met reindeer herders, visited Santa's world, and discovered the magic of the Arctic. From icy hotels and glowing minerals to ancient Sámi traditions and bustling elf workshops, Lapland has shown just how extraordinary the far North can be!

I hope you had fun learning, imagining, and exploring every page. And who knows–maybe one day you'll visit Lapland yourself and say:

✨ "I remember this from my book!"

Until our next adventure... keep exploring, keep creating, and keep wondering about the world!

P.S. If you loved this trip, don't stop here!
Scan the QR code or check out more books in the Junior Globetrotters series–there are so many more places waiting for you!
And if this book made you smile, laugh, or dream...
⭐ Leave a review and share the magic with other explorers!

Hei hei! (That means "Bye-bye!" in Finnish!)
Sylvia

More books from this series:

www.ingramcontent.com/pod-product-compliance
Lightning Source LLC
LaVergne TN
LVHW052018080726
842905LV00047B/1190

* 9 7 8 4 4 9 3 8 2 0 9 1 8 *